F-35 LIGHTNING IIs

★ ★ ★

BY CARLOS ALVAREZ

BELLWETHER MEDIA · MINNEAPOLIS, MN

Are you ready to take it to the extreme?
Torque books thrust you into the action-packed
world of sports, vehicles, and adventure. These books
may include dirt, smoke, fire, and dangerous stunts.
WARNING: read at your own risk.

Library of Congress Cataloging-in-Publication Data

Alvarez, Carlos, 1968-
F-35 Lightning IIs / by Carlos Alvarez.
    p. cm. – (Torque: military machines)
 Includes bibliographical references and index.
 Summary: "Amazing photography accompanies engaging information about the F-35 Lightning
II. The combination of high-interest subject matter and light text is intended for students in
grades 3 through 7"–Provided by publisher.
 ISBN 978-1-60014-332-8 (hardcover : alk. paper)
 1. F-35 (Jet fighter plane)–Juvenile literature. I. Title.
 UG1242.F5A383 2010
 623.74'63–dc22

                                    2009037594

This edition first published in 2010 by Bellwether Media, Inc.

The images in this book are reproduced through the courtesy of: Ted Carlson / Fotodynamics, pp. 8-9;
all other photos courtesy of the United States Department of Defense.

Printed in the United States of America, North Mankato, MN.
010111   1183

CONTENTS

THE F-35 LIGHTNING II IN ACTION

Enemy planes approach a United States Air Force base. A pilot jumps into his F-35A Lightning II. The F-35A streaks into the sky to **intercept** the enemy planes.

The enemy planes fire **missiles** at the F-35A. It dodges the missiles and returns fire. Two of the enemy planes explode. The remaining enemy planes turn around. The enemy pilots know they are no match for the F-35A.

The F-35 is named the Lightning II in honor of the P-38 Lightning, a popular World War II fighter plane.

JOINT STRIKE FIGHTER

The F-35 Lightning II is an advanced fighter plane. It is extremely fast. It has sensors and powerful computers to help pilots in battle. It can attack targets in the air or on the ground.

Plans for the F-35 began in 1996. It was called the Joint Strike Fighter at the time. Several branches of the United States Armed Forces told builder Lockheed Martin what they wanted in the fighter. The F-35 was the result.

The F-35 comes in three types. The F-35A
is the smallest version. It is designed for air
combat. The F-35B is a short take-off and
vertical landing (STOVL) version. It can take
off and land like a helicopter. The F-35C has
folding wings. It can take off from and land
on **aircraft carriers**.

WEAPONS AND FEATURES

The F-35 is loaded with high-tech equipment. It has advanced communications gear and powerful sensors. Computers help pilots fly the plane. They also help **navigate** and fire weapons.

The F-35 is second only to the F-22 Raptor in air-to-air combat capability.

The F-35 has **stealth technology**. The shape of the F-35 and the materials used to make it help hide it from the enemy. These two features make it difficult for the enemy to spot an F-35 on **radar**.

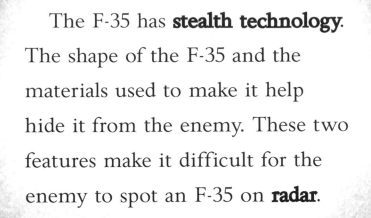

★ **FAST FACT** ★

The U.S. military plans for the F-35 to replace the F-16, A-10, F/A-18, and AV-8B.

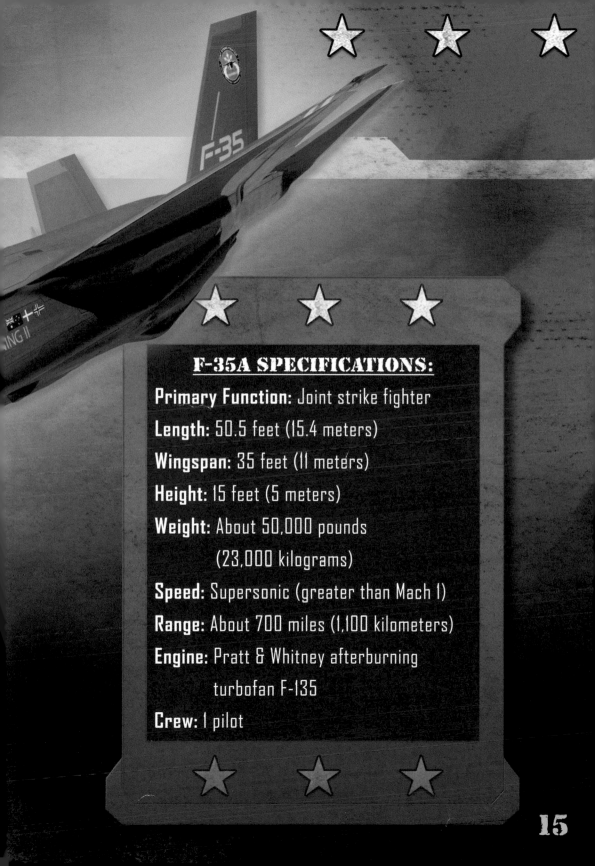

F-35A SPECIFICATIONS:

Primary Function: Joint strike fighter

Length: 50.5 feet (15.4 meters)

Wingspan: 35 feet (11 meters)

Height: 15 feet (5 meters)

Weight: About 50,000 pounds
(23,000 kilograms)

Speed: Supersonic (greater than Mach 1)

Range: About 700 miles (1,100 kilometers)

Engine: Pratt & Whitney afterburning
turbofan F-135

Crew: 1 pilot

The F-35 carries a wide range of weapons. Its main gun is the 25mm GAU-22/A. The F-35 carries its missiles and bombs in two weapons bays. The weapons an F-35 carries depends on its **mission**. The AIM-9 Sidewinder is a heat-seeking missile. It is useful in combat with other airplanes. The joint direct attack munition (JDAM) is a bomb. It can be guided to its target. Pilots use the JDAM to blow up enemy bases and gear.

weapons bay

F-35 MISSIONS

The F-35 can perform a variety of missions. One mission is **air superiority**. The F-35's job is to control the skies and fight enemy aircraft.

The F-35 can also be sent on bombing missions. The F-35's stealth features make it useful as a spy plane. It can fly over enemy territory undetected and gather **intelligence**.

F-35s often work in groups. The pilots are in constant contact with each other. They also talk to their commanders on the ground. The planes can share sensor information with each other. Pilots work together to complete missions and return home safely.

GLOSSARY

air superiority—the ability to counter any force in the air

aircraft carrier—a huge Navy ship from which airplanes can take off and land; an aircraft carrier is like a floating airport.

intelligence—information about an enemy's position, weapons, or movements

intercept—to prevent something, such as a plane or missile, from reaching its target

missile—an explosive launched at targets on the ground or in the air

mission—a military task

navigate—to find one's way in unfamiliar terrain

radar—a sensor system that uses radio waves to locate objects in the air

stealth technology—features that make a plane hard to detect

TO LEARN MORE

AT THE LIBRARY

Alvarez, Carlos. *F/A-18E/F Super Hornets*. Minneapolis, Minn.: Bellwether Media, 2010.

David, Jack. *F-16 Fighting Falcons*. Minneapolis, Minn.: Bellwether Media, 2008.

Zobel, Derek. *United States Air Force*. Minneapolis, Minn.: Bellwether Media, 2008.

ON THE WEB

Learning more about military machines is as easy as 1, 2, 3.

1. Go to www.factsurfer.com.

2. Enter "military machines" into the search box.

3. Click the "Surf" button and you will see a list of related Web sites.

With factsurfer.com, finding more information is just a click away.

INDEX